Very Strange Animals

By Sindy McKay

TREASURE BAY

Introduction
Very Strange Animals

We Both Read books are perfect to read with a buddy—or to read by yourself! If you are reading the book alone, you can read it like any other book. If you are reading with another person, you can take turns reading aloud. When taking turns, it's usually a good idea  for the reader with more experience to read the more difficult parts, marked with a blue dot ●. The reader with less experience can read the parts marked with a red star ★.

Sharing the reading of a book can be a lot of fun, and reading aloud is a great way to improve fluency and expression. If you are reading with someone else, you might also want to take

 the time to talk about what you are reading and what else you know or would like to learn about some of these unusual animals! After reading with someone else, you might even want to experience reading the entire book on your own.

Very Strange Animals

A We Both Read® Chapter Book
Level 3
Blue dot text — Guided Reading Level: S
Red star text — Guided Reading Level: Q

_____

With special thanks to Emma Kocina, Biologist at the California Academy of Sciences, and Rebecca Abrams, M. Ed., for their review of the information in this book

Text Copyright © 2023 by Sindy McKay
Use of photographs provided by iStock, Dreamstime, Shutterstock, and DepositPhotos.
Use of photo of pink fairy armadillo on page 37 courtesy of Earth.com.

All rights reserved

We Both Read® is a trademark of Treasure Bay, Inc.

Published by
Treasure Bay, Inc.
PO Box 519
Roseville, CA 95661 USA

Printed in China

Library of Congress Control Number: 2022942087

ISBN: 978-1-60115-373-9

Visit us online at:
WeBothRead.com

PR-11-23

Table of Contents

Chapter 1
How Strange 2

Chapter 2
It's Prehistoric 4

Chapter 3
Under the Sea 10

Chapter 4
What Is It??? 14

Chapter 5
Hiding in Plain Sight 18

Chapter 6
Eyes, Ears, and Nose 24

Chapter 7
Big and Little 30

Chapter 8
Lovely to Look at 36

Glossary 42

Questions to Ask After Reading 43

CHAPTER 1 HOW STRANGE

Chicken

- How many animal species would you guess there are in the world? One thousand? One hundred thousand? A million? The truth is no one really knows for sure. But scientists estimate between eight and nine million known species. And they believe there are many more yet to be discovered.

  Many species are well known and can be easily seen in homes, backyards, zoos, farms, and parks.

  Others are not as common and seeing them may be rare.

Yellow shaggy frogfish   Komodo dragon

★ You will probably see some animals in this book that you have never seen before.

Some are very odd looking, like the yellow frogfish.

Some may look a bit scary, like the Komodo dragon.

Some are just plain cute, like the tarsier (TAR-see-er).

And some, like the lamprey (LAM-pree), may seem so alien, it's hard to believe they exist here on Earth!

Tarsier   Lamprey

CHAPTER 2   IT'S PREHISTORIC

Cassowary

- Some species of animal—like modern humans—are relatively new. Some are positively prehistoric!

    Take, for example, the cassowary (CASS-e-ware-ee), a colorful bird found only in Australia, Indonesia (In-doe-NEE-ja), and the island of New Guinea (GIN-ee). Like all birds, cassowaries are modern dinosaurs—but cassowaries have more similarities than most birds to some of the ancient two-legged dinosaurs. They were even one of the earliest types of birds to evolve after the large dinosaurs became extinct.

    Cassowaries are more prey than **predator**, but with claws similar to those of a velociraptor (vel-AH-se-rap-tore) dinosaur, they are quite capable of defending themselves when necessary. Their inner claw is particularly lethal and can cut like a five-inch dagger when the bird leaps up high to slash down on its opponent.

Inner claw

Shoebill stork

★ The shoebill stork is another modern-day dinosaur. More closely related to pelicans than storks, these huge birds can grow to be as tall as a person and may have an eight-foot wing span. They can weigh up to 15 pounds. Despite their size, however, these birds are still able to fly.

They are top **predators** in the swamps of eastern Africa, where they live. Although their favorite prey are lungfish, their powerful beaks allow them to sometimes snack on small crocodiles!

Komodo dragon

- The Komodo dragon has roamed Earth for over four million years. It can be up to 10 feet long and most weigh around 150 pounds, though some weigh as much as 300 pounds. These giant lizards have been known to **aggressively** attack and kill humans when they feel threatened. But don't worry, they are only found on a few lightly populated Indonesian islands.

  Komodo dragons are top predators who feast on everything from small rodents to large water buffalo. They have a venomous bite and 60 extremely sharp teeth. Their jaws can unhinge to open wide, allowing them to swallow huge chunks of meat, and their stomach can expand to allow them to eat up to 80% of their own body weight in one meal. When they feel threatened, they can throw up the entire contents of their stomach to lessen their weight and allow them to run faster.

  Komodo dragons are kind of scary but they sure are cool!

Gharial

★ Gharial (GARE-ee-el) is another dinosaur-like creature. It has been around for tens of millions of years.

Related to alligators and crocodiles, gharials can grow to be 20 feet long. That's about as long as 20 cats. And they can weigh as much as 400 pounds. That's *40* cats!

They have over 110 teeth in their long, slender snout. However, they are not **aggressive** and would much rather eat fish than take a bite out of you.

Gharial

Babirusa

- Five-thousand-year-old cave paintings have been found of this odd creature called a babirusa (buh-buh-ROO-sa). The male of this species has four "tusks," which are actually teeth. Two of these teeth break through the skin of their snout and curve up and back toward their eyes. These curved tusks may grow long enough to poke through their skull.

Echidna (uh-KID-na) have porcupine-like spines, a bird-like beak, and a kangaroo-like pouch—and they lay eggs like a reptile! Even though they lay eggs, scientists classify these fascinating animals as mammals because they feed their babies milk. Species of echidna have been around for about 20 million years.

Echidna

Lamprey

★ Tens of millions of years is no time at all for the strange-looking lamprey (LAM-pree). Lampreys, which look a lot like eels, have been around for almost *300 million years*. They are found mainly in the Atlantic Ocean and can range in size from five inches to over three feet long. A jawless fish with a mouth full of little teeth, lampreys attach themselves to fish and suck out their blood.

Even older than lampreys are chambered nautilus (NOT-uh-lis). These alien-looking animals have been around at least 500 million years. They have two eyes, but rely on their sense of smell to find and capture prey in their tentacles. They may not look like they can move very easily, but the nautilus can shoot through the water using jet propulsion.

Chambered nautilus

Eye

Tentacles

CHAPTER 3 UNDER THE SEA

Boxfish shell (carapace)

Boxfish

- Many of the strangest animals on the planet live in Earth's oceans. The boxfish shown above is the only known creature that is shaped like a square (although wombats do have square poop!).

  Instead of a skeleton made of bone, this fish has a hard, square-shaped shell, called a carapace (CARE-uh-pace), that helps protect it from predators. The boxfish can also protect itself by secreting a toxin that poisons the water around it.

  The red lipped batfish not only has an interesting looking mouth but also an interesting way of moving. This fish is not a great swimmer, so it uses its pectoral fins to "walk" on the bottom of the ocean.

Red lipped batfish

← Pectoral fins →

Frogfish

★ Frogfish are another type of fish that can be found "walking" on the ocean floor. There are over 50 species of frogfish and not one of them is related to frogs.

They do, however, all have fins that look a bit like legs, and they use to get them where they need to go.

Frogfish

Blue dragon sea slug Sea slug

- There are over 2,000 species of sea slugs found in both the shallow and the deep water of the ocean. Sea slugs are known for their beautiful bright colors and unique patterns.

  The tiny blue dragon sea slug is barely an inch long and floats along on the ocean's surface in temperate and tropical waters around the world. Its blue coloring makes it hard for predators to see them in the water, but camouflage is not their only defense.

  Blue dragon sea slugs can feed on large, venomous prey, such as Portuguese man o' war, then store that creature's stinging cells in their own bodies for later use. This stored venom remains active even after they die, so humans who pick up or step on them will still feel their painful sting.

Sea slugs

Goblin shark

★ Here's a face you don't want to see when you're going for a swim in the ocean! But don't worry, you probably won't. The goblin shark is a rare species of deep-sea shark. It is sometimes called a living fossil as it is the only remaining member of an otherwise extinct family of sharks.

Believe it or not, the picture below is also a type of shark called a wobbegong (WAA-buh-gong) shark. These sharks have been around for over 11 million years and are sometimes called carpet sharks. Can you guess why?

Wobbegong shark

CHAPTER 4   WHAT IS IT???

Duck-billed platypus

Armadillo girdled lizard

- Duck-billed **platypus** (PLAT-a-puss). Armadillo (arm-a-DILL-o) girdled lizard. Scorpion fly. Elephant shrew. These may sound like animals made up in someone's imagination, but they are all very real. Their names can be confusing though.

    A duck-billed platypus is not a duck.
    Armadillo girdled lizards are not related to armadillos.
    A scorpion fly can't sting.
    And an elephant shrew is much, much smaller than an elephant!
    Still, when you look at these pictures, you can understand how they may have been given their names.

Scorpion fly

Short-eared elephant shrew

Duck-billed platypus

★ So, what is a duck-billed **platypus** if it's not a duck? Like the echidna, they are an ancient type of mammal called monotremes (MAH-no-treems). They are found only in Australia and they were not even discovered until about 200 years ago.

At the time, many people thought that the platypus was a hoax. With webbed feet, a duck-like bill, a tail like a beaver, a body like an otter, and babies that hatch from eggs, people thought someone had just made up this animal from parts of other animals.

Newly hatched duck-billed platypus

15

Armadillo girdled lizard Armadillo

• While the armadillo girdled lizard has the word "armadillo" in its name, some people think it looks more like a tiny three-inch-long dragon! This lizard is not related to armadillos, but it has plates and will roll into a ball to protect itself from predators like its namesake.

A scorpion fly may appear scary, but its scorpion-like tail is just for looks. This insect is in no way related to scorpions. It cannot sting or inject venom and is completely harmless to humans. A main staple of its diet is eating other dead insects that it often steals from spider webs.

Scorpion fly

Scorpion

Elephant shrew

★ Believe it or not, the elephant shrew *is* related to elephants! Well, maybe.... Elephants are in the Afrotherian (af-ro-THEE-ree-en) group of mammals. DNA studies place these little guys in that group as well. There are many species of elephant shrew. The one shown here with the trunk-like nose is the Somali elephant shrew. It uses its long nose to sniff through leaf litter on the ground to find bugs and worms—its favorite food!

Elephant

CHAPTER 5   HIDING IN PLAIN SIGHT

Leaf-tailed gecko

- Survival in the animal world means adapting to your environment, so many animals have evolved to be able to hide in plain sight. The next time you're in a forest, look closer at the trees, leaves, and flowers. You may be surprised to discover that some of what you're looking at is not part of a plant after all.

A closer look at these trees reveals a leaf-tailed gecko (GE-koe), camouflaged as bark. And that strange appendage that looks like a leaf is actually its tail.

These geckos have no eyelids. Instead, there is a transparent covering that protects its eyes. It uses its tongue to keep the covering clean.

This is the potoo

This is wood

This is the potoo

This is a tree branch

★ If you are too big to be a leaf on the tree, maybe you can just imitate an entire branch. The long tailed potoo (poe-TOO) is a bird that does just that. It's hard to tell which part of the tree is bird and which part is wood! During the day, the bird's camouflage helps keep it hidden from monkeys and large birds, such as falcons, that prey on it.

At night, their large eyes and wide mouth help them capture beetles, moths, termites, and more.

Thorn bugs are a favorite snack for many birds, so camouflage is important for them too.

From far away, this bug looks like a sharp green thorn on a plant stem. Up close the bug is colorful and not at all sharp. Most predators, however, don't want to risk it!

Potoo

Thorn bug

Flower mantis

While some animals use camouflage to keep predators away, some use it to lure prey to them. The flower mantis will **disguise** itself as part of the flower it sits on. It stays very still until its prey arrives then quickly snatches it up.

Mantises can be good for a garden because they eat bad bugs like aphids, flies, and locusts. Sadly, they also eat butterflies and ladybugs.

The crab spider can blend in perfectly with a goldenrod flower. Scientists, however, disagree on the purpose of its camouflage. Is it to lure prey to the flower? Or is it to hide from predators? Or could it be for some other reason not yet known? Scientists will keep searching for the answer!

Devil flower mantis

Crab spider

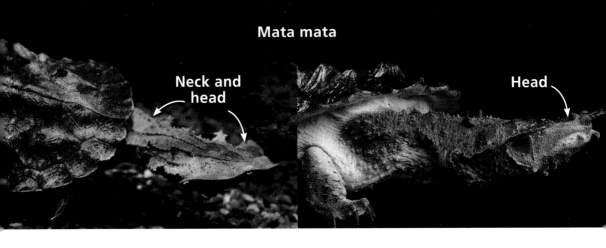

Mata mata

Neck and head

Head

★ The strange mata mata (MAH-tah-MAH-tah) turtle has a long neck with flaps of skin and whiskers. Add to this the bumps and ridges on its shell and this turtle looks like fallen leaves and tree bark floating in the river.

The mata mata turtle's long neck helps when reaching out to grab passing fish, its favorite prey.

The black heron uses a different kind of **disguise** to attract prey. It folds its wings to create a tent-like umbrella over the water. Fish below are attracted to the cool shade it provides and the black heron quickly gobbles them up.

Black heron

Leafy sea dragon

- The leafy sea dragon, a relative of the sea horse, is another undersea master of disguise. Can you tell where the seaweed stops and the animal begins? This disguise helps them easily capture prey, such as shrimp and small fish.

  Their amazing camouflage also keeps them relatively safe from predators...not to mention their sharp, needle-like spines. Ouch!

  Mediterranean octopuses are good at using camouflage to both avoid predators and lure prey. They can change not only the color of their skin but also the texture! It will change from smooth to bumpy and vary in color to closely match the sea floor.

  Octopuses are amazing in so many ways. Take some time to do a little research on them. You won't be disappointed!

Octopus with purple camouflage

Octopus

Stonefish

★ Then there is the stonefish. This is possibly the most dangerous fish in the ocean. The venom in the spines of their back fin can kill an adult human in less than an hour. They are difficult to see so divers may be stung by accident. Luckily there are antivenoms (AN-tee-VEN-oms) available. Stonefish use their venom to keep predators away. Even tiger sharks know better than to mess with them!

Stonefish do not use venom to kill their prey. Instead, they use camouflage to patiently wait for a fish to get close then they quickly attack. They swallow their catch whole in less than a second!

Stonefish

CHAPTER 6   EYES, EARS, AND NOSE

Mantis shrimp

- As animals evolved over millions of years, their sight, hearing, and sense of smell adapted to the environment in which they lived. Some of these adaptations created some very strange and beautiful creatures.

    When it comes to amazing sight, the mantis shrimp (which is neither mantis nor shrimp!) is at the top. Their stalk eyes can move and see independently, and they see a much fuller range of the color spectrum than humans can.

    There are about 450 species of mantis shrimp. All the species have club-like claws that can strike out with the speed of a bullet! The species shown here is what's known as a "smasher." It uses the claws to smash the hard shells of snails and mollusks (MAW-lusks). This strike is strong enough to smash aquarium glass.

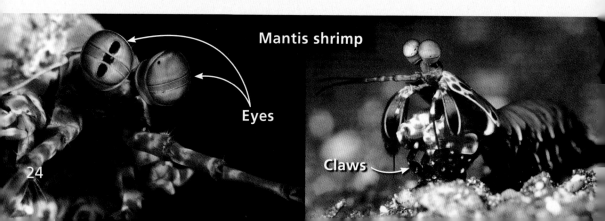

Mantis shrimp

Eyes

Claws

Aye-aye lemur

★   The aye-aye lemur (EYE-eye LEE-mer) seems to have a constant look of shocked surprise. Its bright, wide eyes help the aye-aye hunt in the dark.

It hunts by tapping on a tree and listening to hear if insects start moving under the bark. It then drills a hole with its teeth and uses its long middle finger to reach in and pull out its prey.

Galagos (GAL-a-gos) are adorable primates native to Africa. These little creatures also hunt at night. Their big eyes help them see in low light and their big ears can turn independently to help them hear everything around them. Commonly known as bush babies, their cry sounds much like a human baby. It can be scary if you don't know what it is!

Galago

Natterer's bat

- Speaking of eyes and ears....

There is a common misconception that bats have exceptional hearing but poor eyesight. In truth, they see quite well. However, it is their hearing that is truly awesome and puts human hearing to shame.

There are so many different types of bats that one out of every five mammal species is a bat. The species with the largest ear-to-body ratio may be the tiny leaf-nosed bat. They are only about five inches long and their ears are about one whole inch of that. Like all bats, they use their remarkable ears to hear sounds as slight as the fluttering of an insect's wings.

Leaf-nosed bat

Bats also use their ears for something called echolocation (ek-oh-low-KAY-shun), which allows them to hunt in total darkness. The bat sends out a sound wave, which bounces off an object and returns to let them know where and how far away that object is.

Saiga

★ Animals with the best sense of smell include bears, dogs, sharks, and the all-time best smeller, the elephant.

The animals presented here may not be among the super-smellers of the animal kingdom, but they sure do have interesting noses! The saiga (SIE-guh) antelope, for example, have large noses that hang down over their mouths. They do have an excellent sense of smell, but it is thought that the unique design of their nose is more to help filter out dust in the summer and warm up cold air inhaled in the winter.

There are several species of softshell turtle, and they all have long snouts—or noses—with nostrils at the tips. They are often found buried in mud, sand, and shallow water, and use their snouts like snorkels to break the surface and breathe.

Softshell turtle

Star-nosed mole

- The star-nosed mole may have the strangest nose you'll ever see. Although it may look like a creature from outer space, in reality it spends most of its time underground right here on Earth. Those huge claws make effective shovels for making tunnels!

   This animal has a great sense of smell, but its odd-looking nose is actually more sensitive to *touch* than any other animal on the planet. With eyes that are basically useless, the star-nosed mole uses its incredibly sensitive nose to feel and smell its way around dark tunnels and find prey. Most often found in wet marshlands, this animal is an excellent swimmer. It can use that remarkable nose to sniff out prey *underwater*—one of the very few mammals able to smell in water.

Golden snub-nosed monkey

★   Golden snub-nosed monkeys have hardly any nose at all. Scientists are not really sure why they have these flat, upturned noses. It does not seem to affect their sense of smell. These monkeys sometimes hide their heads between their knees when it rains to keep the water from going up their noses.

We move from almost no nose at all to a nose you can't miss. This is the proboscis (proe-BOSS-kiss) monkey. Scientists are not sure about the purpose of this nose either. But they have noted that the males with the largest noses seem to get the most girlfriends. Beauty is in the eye of the beholder!

Proboscis monkey

CHAPTER 7   BIG AND LITTLE

Saltwater crocodile

- Animals are often broken into six main groups: insects, mammals, birds, **amphibians**, reptiles, and fish. The diversity in size within each of these groups is enormous!

    The largest known reptile is the saltwater crocodile. Some have been as long as 23 feet and have weighed over 2,000 pounds! They have about 66 teeth and perhaps the greatest bite pressure of any animal on Earth.

    At the other end of the reptile family is one of the smallest—the adorable pygmy chameleon. In the wild, these little guys are found primarily in central East Africa. However, they are known to be friendly and don't seem to mind being handled. This makes them popular as pets and they can be found in reptile tanks just about anywhere.

Pygmy chameleon

Giant weta

★ The giant weta (WET-uh) is often considered to be the largest insect. They can weigh almost three ounces—that's about as much as three double-A batteries. Compare that to the weight of a fly that you can barely feel when it lands on your skin!

The world's smallest insects appear to be various species of the fairyfly wasp. Fairyfly wasps can be found all over the world but they are rarely seen or photographed because they are so tiny— about two hundredths of an inch or about the width of an eyelash.

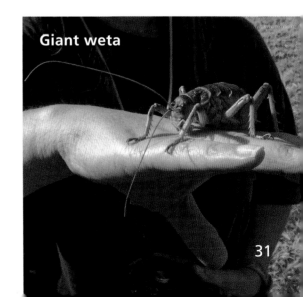

Giant weta

31

Giant salamander

- Most scientists agree that the largest amphibian is the Chinese giant salamander. This ancient species has been around for over 170 million years. It can weigh up to 110 pounds and be up to six feet long. Extremely rare, these animals can still be found in cold mountain streams in Asia.

At the other end of the amphibian size scale is the New Guinea frog. It is currently considered by many to be the smallest amphibian in the world, but that could change as new discoveries are being made every day. Able to spring like crickets, they are not easy to catch!

These tiny frogs eat mites and other prey that are too small to satisfy a bigger frog.

New Guinea frog

Blue whale

★ The largest mammal on land is the African elephant but they can't compete with the largest ocean mammal—the magnificent blue whale. On average, it would take about 30 elephants to equal the weight of one blue whale. They can grow to over 100 feet long. It has a heart as big as a bumper car and its tongue can weigh as much as a small elephant. Most scientists believe they are the largest animals ever to exist—even bigger than dinosaurs!

There are many tiny mammals in the world, including many species of jerboa (jer-BOE-a). The smallest species is the pygmy jerboa. Their bodies are two to three inches long and their tails are about the same length as their bodies. Jerboa have long kangaroo-like legs that allow them to jump great distances and run in a zig-zag pattern to escape predators.

Jerboa

Whale shark

- There are many types of big fish, but none are bigger than the whale shark. It can weigh well over 21 tons and be over 41 feet long. These big guys live in tropical waters and feed on tiny crustaceans (kruh-STAY-shens) like krill and crab larvae (LAAR-vuh). Whale sharks are often found with remora (re-MORE-a)—also called "suckerfish"—attached to them. Remora act like cleaners, eating the parasites on the whale shark's body.

Among the smallest fish in the world are various species of goby (GO-bee). The one shown here is a pink-eyed goby, for obvious reasons. Most goby fish species are less than an inch long.

Pink-eyed goby

Ostrich

★ There is no question that the ostrich (AWE-stritch) is the largest in the bird family. These huge birds can grow to be nine feet tall and can weigh over 250 pounds. That's much taller than even the tallest basketball player.

Some people think these big birds bury their head in the ground to hide. The truth is, they lay their eggs in nests that they dig in the ground and poke their heads in to turn the eggs over several times a day.

The smallest birds are hummingbirds. The brightly colored bee hummingbird is considered to be the smallest bird of all. These tiny birds build tiny nests that are about two inches in diameter. They usually lay one or two eggs at a time and each egg is smaller than a pea.

Hummingbird in nest

Bee hummingbird

CHAPTER 8   LOVELY TO LOOK AT

Red panda

- It has a face like a panda bear and a tail like a raccoon, but the red panda is not closely related to either one. It is actually in a family all of its own—the ailuridae (al-YOUR-i-day.) It's thought to be the last surviving member of this family of mammals. Slightly larger than a domestic cat, red pandas live in trees in the Himalayas and other parts of China.

  When it comes to sweet faces, the axolotl (AX-e-lot-el) is hard to beat. Just look at that smile! Most salamanders are born in water and morph into land creatures, but this critter stays in the water all of its life. Because they are able to grow back lost limbs, these little cuties are often used by scientists for research.

Axolotl

Hairy squat lobster

★ The hairy squat lobster is just one of the many colorful creatures that live under the sea. Less than half an inch in size, this little pink beauty is also called a fairy crab.

Speaking of fairies, the strange little mammal below is called a pink fairy armadillo. It is less than five inches long and is the smallest type of armadillo. Found only in Argentina, it spends most of its life underground. Like other armadillos, it has a shell or carapace (CARE-a-pace). The pink fairy armadillo's carapace is softer and more flexible than most armadillos. It only covers about half of its body. The pink color comes from blood vessels close to the surface. Those big claws are used to dig through the sandy plains and deserts where it lives.

Use of photo courtesy of Earth.com.

Pink fairy armadillo

Lilac-breasted roller

Curly-crested aracari

Golden pheasant

Painted bunting

• Some of the most vivid colors in the animal kingdom are found in birds. Just look at the gorgeous colors represented here—and check out the cool feathery fashions on the heads of some of these birds!

The colorful Mandarin duck is native to China and Japan. Because these ducks stay with one partner for life, they are often considered a symbol of love.

The Luzon bleeding heart pigeon looks like it is bleeding from its chest. But don't worry, it's just the bird's natural coloring.

Mandarin duck

Luzon bleeding heart pigeon

Secretary bird

★ When it comes to glamour, the secretary bird wins, hands down. Check out those eyelashes and long legs. This bird looks like it's strutting on a high fashion runway!

Maybe it's all that strutting that gives this bird one of the strongest and fastest kicks in the animal kingdom. This strong, fast kick comes in handy when hunting down their favorite prey—snakes.

Secretary bird

Snake being caught by secretary bird

Tarsier   Bald uakari

• The secretary bird may be glamorous, but the adorable tarsier (TAR-see-er) wins the prize for cuteness. Tarsiers are only four to five inches long and weigh less than half a pound. They are one of the world's smallest primates. It's hard to believe they are in the same family as a gorilla!

Bald uakari (wa-KAR-ee) are friendly primates with hairless, red faces that allow their facial expressions to be seen more clearly than other primates.

What emotions do you see in these faces?

Bald uakari

★   The incredible diversity of animals on Earth is awe inspiring. The animals presented in this book make up just a tiny percentage of the living species. There are so many more to be explored and discovered. The more we know, the better we can help to keep all their populations thriving.

The world belongs to all the creatures on it. We must learn to share with care.

Glossary

amphibian
cold-blooded animals with backbones and no scales that live part of their lives in water and part on land

crustaceans (kruh-STAY-shens)
the group of mostly water animals with a body made of segments, a tough outer shell, two pairs of antennae, and limbs that are jointed, such as crab, lobsters, and shrimp

hoax
something fake that is passed off as real

mollusks
a kind of animal with a soft body and no backbone that usually lives in a shell, such as snails and clams

primate
any mammal of the group that includes lemurs, lorises, tarsiers, monkeys, apes, and humans

proboscis (proe-BOSS-kiss or proe-BAH-sis)
a long, flexible snout, as in the trunk of an elephant, or a tubular feeding organ, as that of a butterfly

venom
poison produced by an animal and passed to a victim by biting or stinging

Questions to Ask after Reading

Add to the benefits of reading this book by discussing answers to these questions. Also consider discussing a few of your own questions.

1 There are many unusual animals in this book. Which one would you most like to see in the wild or in a zoo? What do you find interesting about that animal?

2 The environment affects the way animals evolve. What are some of the differences you might see in animals native to areas where it snows almost all year long and those animals native to hot jungles?

3 Which animal in this book would you like to learn more about? How do you think you could learn more about it?

4 Some people try to keep exotic wild animals as pets. Do you think that is a good idea? Why or why not?

If you liked **Very Strange Animals** here are some other
We Both Read® books you are sure to enjoy!

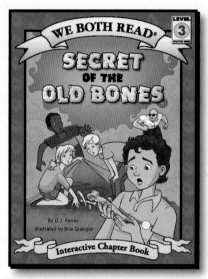

You can see all the We Both Read books that are
available at WeBothRead.com.